I0489814

GUIDE PRACTICAL
HANDBOOK FOR
MOBILE DEVICE
REPAIRS

How to repair and service cell phone and systems

Scarlett R. Lewis

Table of Contents

CHAPTER ONE

INTRODUCTION

How to repair and service mobile devices, is simply outlines and complete analysis of system maintenance process. To be a mobile repairer of devices required basic attitude to have cordially relationship with your client.

Replacement LCD Display Screen Of A Mobile Phone

Dissemble first

To trade a connector-type LCD Display Screen is very easy. Just disassemble the telephone and dispose of the connector of the display. Replace the erroneous show display screen with a new one by way of the use of inserting the connector to the PCB. Disassemble the Smartphone the use of precision screwdriver and cell opening tool. You will see the exhibit soldered to the PCB of the Mobile Phone.

Solder joints

Apply flux to the solder joints. Using soldering iron warmness all the solder joints, Be cautious even as the use of soldering iron. Using tweezers or hand, gently pull out the exhibit as quickly as the solder starts melting.

Desolder and connect the necessary wire

Using Desoldering wire and soldering iron, take away all the greater solder paste from the PCB Track. Apply flux to the tune on the PCB and the New Display. Using soldering iron, exercise solder to every the exhibit and the PCB track. Place the exhibit on the PCB and solder each component of the exhibit to the music one with the aid of capability of one. Try to solder every the ends first and then solder the core part.

CHAPTER TWO

HOW TO CONNECT AND FIX DAMAGE DEVICES

> ➤ **Steps in maintaining good communications in mobile line service**

Cleanness and neat environment

Always keep the work area neat and clean, it is utmost essential that you keep your administrative middle neat and convenient due to the truth a dirty and unorganized administrative middle shows on the persona and manners of the man or woman and a clean environment always attracts people. An effortless laptop will typically furnish greater results. If you wish to expand your output, then make positive you keep your region of work very loads neat and clean.

Proper arrangement of tools

Whenever you trade the screws after fixing the phone, make sure that you alternate all the screws in their respective slots. If per danger due to your carelessness you can additionally have misplaced some screws make positive you use screws of equal type, size and structure or else you will simply harm the threads in the slots or the slot itself.

Handle devices with care

You ought to continuously take into account the manner in which you had dismantled the telephone Smartphone due to the truth when you commence reassembling the same, you will have to go exactly in the reverse manner or else you will bypass over something and unnecessarily waste your treasured time.

Earlier it used to be kind of difficult as we had all sorts of handsets such as flippers and sliders. But nowadays, handsets are very lots a piece of cake for disassembling and reassembling. Also while disassembling and assembling, you ought to be extra cautious about the LCD as it is the most sophisticated part and the most luxurious part, in a phone.

Have essential and necessary appliance

When you take a job to restoration a cell phone make positive that you have all the suitable elements to replace. It would be a acquire if you keep the spare elements of the most widespread fashions in stock so that when the handsets come for repairs, you can repair them proper now barring dropping any time.

Secure client properties under your custody

Keep the SIM card and memory card of the purchaser safely, when you get hold of any handset for repairs from a customer, continuously make it an element to preserve the SIM and memory card safely. It would be greater if you do this in an outfitted manner such as maintain it in a separate small container with the customer's title and or model of the handset or the volume of the receipt which you might also additionally have issued to the customer.

Avoid excuses while discharging duties

When you fix a phone, make positive that you do no longer delete the files till required. The facts is a complete lot greater fundamental than the handset itself. If it truly necessary, take a backup of the data and it is utmost quintessential to take the customer's consent formerly than deleting facts or performing a manufacturing facility reset.

> ➤ **Multimeter And DC Usage**

Guide in Motor Mobile Phone

To take appear at vibrator or motor of a mobile phone, maintain the multimeter in Buzzer Mode and take a look at the vibrator. Value ought to be 7 to 17Ohms. If the value is between 7-17 Ohms then the vibrator is good. Otherwise change it.

Guide in Ringer Mobile Phone

To take seem as if the ringer of a mobile Smartphone is inaccurate or damaged, keep the multimeter in buzzer mode and take seem at the ringer. Value wants to be between 7 to 10 Ohms. If the charge is between this fluctuate then the ringer is suitable and does now no longer favor replacement. If the fee on multimeter is 4-5 or 12-14 then alternate the ringer.

Guide in Earpiece Mobile Phone

Check the speaker or earpiece with a multimeter on Buzzer mode. Value ought to be in several of 24 to 36 Ohms. If the charge is in this range then the speaker or earpiece is good enough and prefer no longer be changed. Otherwise, change the speaker earpiece.

Guide to Microphone

Keep the multimeter in buzzer mode and take a look at the microphone. Value inspecting on the multimeter have to be in fluctuate of 600 to 1850 Ohms. There will moreover be a Beep or Buzz sound from the multimeter.

Guide to Keypad Mobile Phone

Keep the multimeter on Buzzer mode and check Rows and Columns or the Key Pad. If there is Beep or Buzz sound from the multimeter then Keypad is ok, in any different case it is faulty.

Guide to Mobile Phone Electronic unit

Check the SMD coil with a multimeter on Buzzer Mode. If it is authentic then it the multimeter will provide a Beep or Buzz sound. If there is no sound then the coil is faulty. Replace it with a new one.

Guide to Mobile Phone Resistor

Check it with a multimeter on Buzzer Mode. If it is authentic then the multimeter will supply a Beep or Buzz sound. If there is no sound then the SMD resistor is faulty. Replace it with a new one.

Guide to Mobile Phone Capacitor

Test SMD Capacitor with a multimeter on Buzzer Mode. If it is perfect then the multimeter will NOT grant any Beep or Buzz sound. If there is sound then the capacitor is faulty. Replace it with a new one.

Guide to Mobile Phone LED

Keep the multimeter in Buzzer mode and take a look at the LED. If the LED is desirable then they will glow in any different case not.

Guide to Mobile Phone Coil Boosting

Check for continuity. If there is continuity then the coil or the

Boost Coil is proper in any different case it is faulty.

Guide to Mobile Phone IC

Use an Analog DC Power Supply to check Network IC. Switch ON DC Power Supply and title any volume from your phone. The Needle of the DC Ampere will commence moving. This suggests that the Network IC is good enough and no longer fault.

Guide to Mobile Phone Power IC

Adjust voltage of the DC Power Supply to 4.2. Place the Red Probe or Test Lead of the DC Power

Supply to the + of the Battery Connector of the telephone Smartphone and the Black Probe or Test Lead to −: If DC Ampere is over 6 then Power IC or CPU is damaged. Check by using the use of altering Power IC and the CPU one by using the usage of one.

If there is no action of the Ampere Needle of the Power Supply then the Battery connector, On / OFF Switch Track, RTC or Network Crystal is damaged. Give warmness to these elements the utilization of heat air blower. If the trouble is no longer solved then with the resource of alter them one via ability of one.

If the Ampere Needle fluctuates beneath two ten there need to be hassle with software program software or RTC Real Time Clock. If the Ampere needle stands at some regular thing then there is problem with the Flash IC. If there is beep sound from the DC Power Supply then there is problem with + and − or the telephone handset is short.

When checking an erroneous mobile phone with DC Power supply, be a part of the Red Probe to + and Black Probe to − of the Battery Connector of the Mobile Phone. Most telephone restores human beings and technician check entirely above aspects to remedy cell hardware problems. All specific aspects consisting of SMD digital factors and ICs are usually no longer checked for fault. There is no right sure test for these parts. The bother is each solved thru jumper or thru trial and error.

CHAPTER THREE

TROUBLESHOOTING DEVICES

Common issues in repairing devices

Issue of Storage Space

Limitless indoors storage to preserve higher HD videos, takes greater pictures, down load greater music, and does many things. That's, however, now not continuously the trouble in view that most people are plagued with the Storage residence taking walks out issue at a stunning time.

Treatment requirement

Install Files with the resource of Google or SD Maid from the Play Store and use it to scan and get rid of vain statistics ingesting up your storage. Delete all significant videos, music, pictures, and archives you no longer need; they take the most storage space.

If you have many photos, assume about backing them up to Google Photos, and deleting their furnish files from your phone. Lastly, you may also moreover get an SD card and bypass most of your archives to it

Issue of Battery capacity

Phone's battery existence boil down to three things. Capacity, energy efficiency, and how you use your phone. And while you have fewer controls over the first two, you can alter your utilization to have greater battery life on your phone.

Treatment requirement

Manually reduce your show display brightness or enable Adaptive Brightness to make positive you're no longer the use of greater power on the show when no longer needed. If your Smartphone has a higher refresh rate, you may also additionally want to set it to Adaptive refresh rate or pick out a reduce refresh rate. Turn on darkish mode on AMOLED indicates to decrease show display battery usage. Charge you're mobile properly.

Issue of Overheating

It's everyday for your cell phone to get warmness while the utilization of it. But as quickly as it is getting hot, there is greater bother to it, and you have to restoration it as hastily as possible. Overheating is one of the familiar troubles that can show up on any phone, relying on what the character is doing.

Treatment requirement

Avoid the utilization of your Smartphone even as charging it to stop it from getting hot. Also, make positive there's enough air drift or cooling the location you feel your phone.

If your phone Smartphone overheats while taking section in games, the use of a special app, or on foot some intensive tasks, replicate on consideration on doing away with its cowl and imparting a cooler atmosphere. Take breaks between hardcore gaming durations to cool down your phone and keep your battery. Avoid the utilization of your phones for too prolonged under immoderate temperatures to preserve away from overheating.

Issue of Charging

Clean your charging port, Confirm your charger and cable are working. Confirm you've associated to a working power source. Restart your phone

Issues of Network

Having a gradual neighborhood connection when you favor it the most can be frustrating; alternatively you mustn't get mad due to the truth it is clearly one of the regular troubles we face as Smartphone users.

Treatment requirement

Turn on Airplane mode on your Smartphone for about 35 seconds, and then flip it off again. Switch to each and every different Wi-Fi neighborhood if available. If you are on a mobile community make positive you are the use of the relevant net configuration. Install the contemporary mannequin of your OS.

Issues of slow operating

We all pick to get our stuff carried out barring issues so we can get to the subsequent one. However, it's now not continuously as easy as that, specifically if your phone's hardware isn't that vast.

Treatment requirement

Delete apps you're no longer using, make certain you have ample storage space. Reduce your records apps and tasks, Update your walking system. Scan your telephone for viruses and delete junk files. Reduce your animation tempo with the aid of Developer Options. Turn off place issuer even as now not in use.

Issue of App Crash

If you have a specific app it is consistently crashing, or your

Smartphone itself tends to freeze sometimes,

Treatment requirement

Update the referred to app from the Play Store or your respective app store. If it's though now not working, resort to the app. Locate the app on your domestic screen. Long press on it, then faucet App facts or Select Storage, then clear cache and data. Also, make certain you have adequate memory and storage location on your phone, and then change your Smartphone software.

Issue of not powering

Arguably one of the greatest Smartphone problems is when value would possibly no longer or flip on.

Treatment requirement

But first, you ought to confirm the issue isn't from your charger, charging port, or something related. Cell however won't value or flip on at the case.

www.ingramcontent.com/pod-product-compliance
Lightning Source LLC
Chambersburg PA
CBHW070522220526
45467CB00002B/805